Understanding Acid-Base Disturbances

Understanding Acid-Base Disturbances

A Simplified Problem-solving Approach

Manuel A. Castro, MD

To order additional copies of this book, contact:
Xlibris Corporation
1-888-795-4274
www.Xlibris.com
Orders@Xlibris.com
68134

Contents

Introduction to Acid-Base

The pH of the body is maintained at 7.40. This is accomplished by two main mechanisms which balance each other constantly—CO_2 as well as bicarbonate or HCO_3. Each is regulated mainly by two organ systems—one is the pulmonary system for CO_2 and the other is the kidneys for the bicarbonate or HCO_3.

The lungs regulate the pH by increasing or decreasing the rate with which a person breathes and this regulates the amount of CO_2 in the blood. By breathing fast, more CO_2 will be eliminated from the body and by breathing slowly less CO_2 will be eliminated. An excess of CO_2 will increase the acidity of the blood, and decreasing the amount of CO_2 will decrease the acidity of the blood. The lungs can adjust the pH very quickly, all it takes is for the person to breathe fast or slow to cause this regulatory effect to take place. Therefore, breathing fast will cause an alkalosis and breathing slow, will cause an acidosis. These changes are both due to the pulmonary system so they are called pulmonary acidosis or alkalosis.

The kidneys regulate the pH by retaining HCO_3 from being eliminated in the urine or by eliminating HCO_3 in the urine, as well as other substances. Retaining HCO_3 will increase the amount of base in the blood making it more alkaline and eliminating more HCO_3 reduces the amount of base in the body making it more acid. Therefore, if the HCO_3 is increased the person will develop a metabolic alkalosis and if it is decreased the patient will develop a metabolic acidosis. This regulatory system is slower than the pulmonary system and takes days to fully regulate any type of disturbance.

The amount of hydrogen ions in the body is usually calculated with the following equation:

$$[H+] = 24 \times pCO_2/HCO_3$$

So anything that increases the pCO_2 will cause the amount of hydrogen ions to increase causing an acidosis, and anything that increases the amount of HCO_3 will cause the amount of hydrogen ions to decrease causing an alkalosis. Conversely, anything that decreases the pCO_2 will cause the amount of hydrogen ions to decrease causing an alkalosis, and anything that decreases the amount of HCO_3 in the blood will cause the amount of hydrogen ions to increase causing an acidosis.

With that said, a person can have different types of acid-base disturbance which are:

- Metabolic acidosis: due to decrease in HCO_3
- Metabolic alkalosis: due to increase in HCO_3
- Respiratory acidosis: due to increase in pCO_2
- Respiratory alkalosis: due to decrease in pCO_2

How do we suspect an acid-base disturbance in a patient? It's very simple. The patient will be breathing slow, fast or erratic. The respiration of the patient will be the first sign that you will have of an acid-base disorder. So, in any patient that is breathing fast, slow or erratic you should always evaluate the arterial blood gases, as well as, electrolytes in the blood. It would also help to have urine electrolyte to further assess the acid-base disorder of the patient.

The first step in the evaluation of a patient with an acid-base disturbance is to figure out the primary acid-base disturbance. First look at the arterial blood gases, and see what the pH is. If the pH is less than 7.40 consider an acidosis and if the pH if greater than 7.40 consider an alkalosis. This is overall the predominant disturbance in the patient. Now, the next thing you should do is to look at the pCO_2 of the arterial blood gases. If the pH is less than 7.40 or an acidosis is present and the pCO_2 is high, then the primary disorder in the patient is a respiratory acidosis. But if the pCO_2 is low then the primary disorder cannot be a respiratory disorder. It must be a metabolic disorder, and since the pH is less than 7.40 it has to be a metabolic acidosis, which usually is compensated by a respiratory alkalosis. Now, if the pH is greater than 7.40 and the pCO_2 is low or less than 40 then the primary disorder is an alkalosis and since the pCO_2 is low it is a respiratory alkalosis. But if the pH is greater than 7.40 and the pCO_2 is more than 40 which would cause the pH to go down, then the primary disturbance is a metabolic alkalosis and is likely being compensated by a respiratory acidosis.

We will introduce acid-base disorders first with simple cases, and then go to more complex cases, including mixed acid-base disturbance.

Case 1:

A forty-two-year-old male presents to the emergency room with severe abdominal pain. The patient's respiratory rate is 29 breaths per minute. You obtain the following arterial blood gases: pH 7.49, pCO_2 30, pO_2 99. What is the primary acid-base disturbance of the patient?

    a) Metabolic acidosis
    b) Metabolic alkalosis
    c) Respiratory acidosis
    d) Respiratory alkalosis

By looking at this patient's pH you know it is greater than 7.40. Therefore, the patient's main acid-base disturbance is an alkalosis. Then you look at the pCO_2 and you notice that it is low. If the pCO_2 is low it will cause an alkalosis, which will explain the reason why this patient's pH is alkalotic. Therefore, the primary acid-base disturbance is a respiratory alkalosis since this disturbance is explained by the shift in the pCO_2. So, this patient likely has a respiratory alkalosis due to breathing fast, which is secondary to the abdominal pain that he has. So, once his pain is controlled, the acid-base disturbance will likely be corrected.

Case 2:

A seventeen-year-old female with type 1 diabetes presents with an elevated blood glucose of 420. You obtain arterial blood gases which are as follows: pH of 7.25, pCO_2 of 26, and pO_2 of 99. What is the primary acid-base disturbance in this patient?

    a) Metabolic acidosis
    b) Metabolic alkalosis
    c) Respiratory acidosis
    d) Respiratory alkalosis

By looking at the pH in this patient you notice that it is less than 7.40, which indicates that this patient has an acidosis. Then, you look at the pCO_2

which is low or less than 40. Since this would cause an alkalosis, the primary disturbance is not related to the low pCO_2 but likely to a metabolic disorder which likely is being compensated to a certain degree with a respiratory alkalosis. Knowing that the patient is a type 1 diabetic, she is prone to diabetic ketoacidosis which is likely to be causing this patient's problem.

Case 3:

A twenty-two-year-old male who was admitted due to a sickle cell crisis was started on morphine for the pain. The arterial blood gases were obtained when the patient started to become confused. The readings are: pH 7.25, pCO_2 75, and pO_2 65. What is the primary acid-base disturbance in this patient?

> a) *Metabolic acidosis*
> b) *Metabolic alkalosis*
> c) *Respiratory acidosis*
> d) *Respiratory alkalosis*

As you can see the pH in this patient is less than 7.40, which indicates the presence of an acidosis in this patient. Now, look at the pCO_2 which is also elevated. If the pCO_2 goes up it causes an acidosis which explains the cause of this patient's low pH. Therefore, it is fair to say that this patient has a primary respiratory acidosis since the shift in pCO_2 can explain the shift in the pH.

Case 4:

A forty-one-year-old female was admitted to a hospital due to severe nausea and vomiting, which has not resolved in the last few days. The patient continues to vomit and you obtain the following arterial blood gases: pH of 7.46, pCO_2 of 46, and pO_2 of 99. What is the primary acid-base disturbance in this patient?

> a) *Metabolic acidosis*
> b) *Metabolic alkalosis*
> c) *Respiratory acidosis*
> d) *Respiratory alkalosis*

As you can see in this patient the pH is greater than 7.40, which indicates an alkalosis. The pCO_2 of this patient is high, which would not explain the alkalosis. Since a high pCO_2 will cause an acidosis, most likely this patient has a primary metabolic alkalosis. The patient's vomiting causes excess hydrochloric acid to be removed from the stomach leading to this patient's alkalosis. So the correct answer to this question is metabolic alkalosis.

Compensatory Mechanisms
of Metabolic Disturbances

So far you learned to identify the primary acid-base disturbance by looking at the pH as well as the pCO_2. The human body will compensate for any acid-base disturbance by trying to bring the pH towards normal. If you have a metabolic acidosis or alkalosis, your lungs will try to compensate. The respiratory rate increases to create a respiratory alkalosis to compensate for a metabolic acidosis. The respiratory rate decreases by breathing slower to compensate for a metabolic alkalosis. This will bring your pH as close to 7.40 as possible. If you have a respiratory disorder you will compensate with a metabolic component. That is, if you have a respiratory acidosis or alkalosis, you will compensate by either increasing or decreasing the amount of HCO_3 to maintain the pH as close to 7.40 as possible.

I mentioned earlier that the respiratory system will compensate very fast for any metabolic disorder, since all you have to do is breath fast or slow, but the respiratory disturbances are not compensated as fast as the metabolic ones and can take up to several days. Let's look again at some cases to see if they are compensated adequately. This is the second step in the evaluation of an acid-base disturbance. *After determining the primary acid-base disorder, the next step is to determine if it is well compensated.* If it is not adequately compensated it is because the patient has a secondary acid-base disturbance such as a respiratory acidosis or alkalosis, on top of their metabolic disturbance.

First let's see the compensatory formulas for the expected pCO_2 in metabolic disorders.

- **Metabolic Alkalosis**
  pCO_2 = (Measured HCO_3 - Normal HCO_3) × 0.7 + 40
- **Metabolic Acidosis**
  pCO_2 = (1.5) HCO_3 + 8 (plus or minus 2)

Case 1: Compensation in Metabolic Alkalosis:

A thirty-three-year-old male was admitted to a hospital due to severe gastroenteritis, predominately vomiting, which has been worsening. Arterial blood gases obtained is as follows: pH of 7.46, pCO_2 of 46, and pO_2 of 99. And, the patient's electrolytes were reported as: sodium of 140, potassium of 3.8, chloride of 102, and bicarbonate of 32. What is the primary acid-base disturbance in this patient?

a) Metabolic acidosis
b) Metabolic alkalosis
c) Respiratory acidosis
d) Respiratory alkalosis

We saw earlier that this is a primary metabolic alkalosis and this is further confirmed by an increase in the bicarbonate in the electrolyte of the patient. This metabolic alkalosis is due to the vomiting the patient is presenting with. Is this metabolic alkalosis adequately compensated?

Metabolic alkalosis is caused by loss of hydrogen ions such as in vomiting, where the patient loses excess hydrochloric acid or less commonly by addition of bicarbonate. Usually metabolic alkalosis is corrected by elimination of renal bicarbonate. Other causes of increased renal reabsorption of bicarbonate are due to volume contraction, low potassium, elevated mineralocorticoid activity and chloride depletion. The next thing you need to do now is to calculate if the metabolic alkalosis is adequately compensated. In metabolic disorder, the math works in our favor when the metabolic disorder is fully compensated. Usually, if you look at the pH given to you, which is 7.46, and the pCO_2 of 46, the last two digits of the pH or 46 is within one or two numbers of the pCO_2 which is 46, in a fully compensated metabolic disorder. Now, you would like to confirm this with the compensatory formula for metabolic alkalosis, which is:

$$pCO_2 = (\text{Measured } HCO_3 - \text{Normal } HCO_3) \times 0.7$$

$$pCO_2 = (32 - 24) * 0.7 + 40 = 45.6$$

Now, if the pCO_2 would have been less than 43.6 then this would be a metabolic alkalosis with an associated respiratory alkalosis as well. If the pCO_2 would have been greater than 47.6 then this would have been a metabolic alkalosis with a respiratory acidosis as well. This trick works with both metabolic alkalosis and metabolic acidosis.

Now, if this case was presented as below:

A thirty-three-year-old male was admitted to a hospital due to severe gastroenteritis, predominately vomiting, which has been worsening. Arterial blood gases obtained is as follows: pH of 7.52, pCO_2 of 45, pO_2 of 99, and the patient's electrolytes were reported as: sodium of 140, potassium of 3.8, chloride of 102, and bicarbonate of 37. What is the primary acid-base disturbance in this patient?

In this case as you can see that the last two digits of the pH (52) is not within one or two numbers of the pCO_2 which is (45), and because the pCO_2 is lower than what is expected there must be a added-on respiratory alkalosis on this patient, who already has a metabolic alkalosis. So, if this would have been the case the patient has a metabolic alkalosis and a respiratory alkalosis as well.

Now, if this case was presented as below:

A thirty-three-year-old male was admitted to a hospital due to severe gastroenteritis, predominately vomiting, which has been worsening. Arterial blood gases obtained is as follows: pH of 7.45, pCO_2 of 57, and pO_2 of 99, and the patient's electrolytes were reported as: sodium of 140, potassium of 3.8, chloride of 102, and bicarbonate of 32. What is the primary acid-base disturbance in this patient?

In this case, as you can expect, the pCO_2 should have been 45 or close to 45, but this pCO_2 of 57 is much higher than what is expected, indicating that this patient has a metabolic alkalosis as well as a respiratory acidosis.

Case 2: Compensation in Metabolic Acidosis:

A twenty-three-year-old male alcoholic presents to the emergency room disoriented. You obtain arterial blood gases which are as follows: pH of 7.30, pCO_2 of 31, and pO_2 of 99; electrolytes obtained on the patient shows a sodium of 142, a potassium of 4.2, a chloride of 85, and a bicarbonate of 15. What is the primary acid-base disturbance in this patient?

    a) *Metabolic acidosis*
    b) *Metabolic alkalosis*
    c) *Respiratory acidosis*
    d) *Respiratory alkalosis*

You know that the pH in this patient is acidotic and that the pCO_2 is low. This indicates that this acidosis is not due to a respiratory disorder; otherwise, the pCO_2 would have been elevated or greater than 40. This is reinforced by the fact that the bicarbonate in the electrolytes was low at 15, which definitely indicates that this is a primary metabolic acidosis. Now, it is important to continue evaluating metabolic acidosis (See the Chapter on Metabolic Acidosis Assessment). During this section, we will see if this metabolic acidosis is well-compensated. This is done by checking the last two digits of the pH and the pCO_2. As you can see both are within one to two numbers from each other. If you wish to confirm this, you can do the math with the formula given below:

$$pCO_2 = (1.5) \, HCO_3 + 8 \text{ (plus or minus 2)}$$
$$pCO_2 = (1.5)15 + 8 = 22.5 + 8 = 30.5$$

So, this confirms that the pCO_2 should be close to 30. The pCO_2 in this case was 31, therefore, this is a well-compensated metabolic acidosis.

Now, if this case was presented as below:

A twenty-three-year-old male alcoholic presents to the emergency room disoriented. You obtain arterial blood gases which are as follows: pH 7.30 pCO_2 of 38 pO_2 of 99; electrolytes obtained on the patient show a sodium of 142, a potassium of 4.2, a chloride of 85, and a bicarbonate of 15. What is the primary acid-base disturbance in this patient?

In this case, you can tell that the pH is acidotic at 7.30 and the pCO_2 is less than 40. This must be a primary metabolic acidosis since a respiratory acidosis would have a pCO_2 greater than 40. But when we look at the compensatory mechanism we know that what we expect the pCO_2 to be around 30 or within one or two numbers from 30. But, it is much higher indicating that this patient also has a respiratory acidosis added on to the metabolic acidosis as well or a double acidosis.

Now if this case was presented as below:

A twenty-three-year-old male alcoholic presents to the emergency room disoriented. You obtain arterial blood gases which are as follows: pH of 7.30, pCO₂ of 25, and pO₂ of 99; electrolytes obtained on the patient show a sodium of 142, a potassium of 4.2, a chloride of 85, and a bicarbonate of 15. What is the primary acid-base disturbance in this patient?

As you can see in this case, the pH is acidotic indicating an acidosis as the primary disturbance and the expected pCO_2 would have been around 30 or within one to two numbers from it from 28 to 32. But, it is actually much lower than that (25) indicating that this is a mixed metabolic acidosis and a respiratory alkalosis.

Now let's look at a special case below:

A twenty-three-year-old male alcoholic presents to the emergency room disoriented. You obtain arterial blood gases which are as follows: pH of 7.30, pCO₂ of 31, and pO₂ of 82; electrolytes obtained on the patient show a sodium of 129, a potassium of 4.9, a chloride of 103, a bicarbonate of 15, and albumin of 2 g/dL. What is the most likely diagnosis in this clinical case?

a) *Mixed non anion gap metabolic acidosis and respiratory alkalosis*
b) *Mixed non anion gap metabolic acidosis and respiratory acidosis*
c) *Mixed anion gap metabolic acidosis and respiratory alkalosis*
d) *Mixed anion gap metabolic acidosis and respiratory acidosis*
e) *Normal anion gap metabolic acidosis compensated*
f) *High anion gap metabolic acidosis compensated*

This case has something unique—if you don't take it into consideration you would get the wrong answer. The point is that the albumin is low. *The anion gap on a patient with a low albumin is falsely low for every 1 g/dL of albumin below the normal, which is usually 4 g/dL and you must add 2.5 to the anion gap.* This patient's albumin is 2 g/dL, which is 2 below the normal, therefore, the anion gap in this patient is actually 5 above what it is actually measured. Initially, the anion gap in this patient appeared to be normal at 11. But, after noticing that the albumin is low you add on about 5 points due to the low albumin, which actually brings the anion gap to about 16, which is considered a high anion gap metabolic acidosis. We know that the compensatory mechanism is adequate and that there is no added respiratory disturbance to this case since the last two digits of the pH is very close to the pCO_2. If you do not account for the low albumin, your answer would be a normal anion gap metabolic acidosis compensated and it is actually a high anion gap metabolic acidosis well-compensated.

Compensatory Mechanisms
of Respiratory Disturbances

So far you learned to identify the primary acid-base disturbance by looking at the pH as well as the pCO_2. The human body will compensate for any acid-base disturbance in trying to bring the pH towards normal as we mentioned in the previous chapter. If you have a respiratory disorder, you will compensate with a metabolic component. That is, if you have a respiratory acidosis or alkalosis, you will compensate by either increasing or decreasing the amount of HCO_3 to maintain the pH as close to 7.40 as possible. If there is a primary respiratory acidosis, the HCO_3 will increase to compensate and if there is a respiratory alkalosis, the HCO_3 will decrease to compensate.

I mentioned earlier that the respiratory system will compensate very fast for any metabolic disorder, since all you have to do is to breath fast or slow, but the respiratory disturbances are not compensated as fast as the metabolic ones and can take several days. So, if a patient has a respiratory disturbance it is important to determine if the disturbance is acute, lasting for just a few hours or chronic, lasting for some days. This is because the compensatory mechanism expected for acute respiratory disturbance is different from the compensatory mechanism for the chronic disturbances. Let's look again at some cases to see if they are compensated adequately. This is the second step in the evaluation of an acid-base disturbance. *After determining the primary acid-base disorder, you need to determine if it is well compensated.* If it is not adequately compensated, it is because the patient has a secondary acid-base disturbance such as a respiratory acidosis or alkalosis on top of their metabolic disturbance.

First let's show the formula for the expected HCO_3 for respiratory acidosis and alkalosis depending on whether it is acute or chronic:

- **Acute Respiratory Acidosis**
  1 meq/L for every 10 mmHg increase in pCO_2

- **Chronic Respiratory Acidosis**
  3.5-4 meq/L for every 10 mmHg increase in PCO_2

- **Acute Respiratory Alkalosis**
  2 meq/L for every 10 mmHg decrease in pCO_2

- **Chronic Respiratory Alkalosis**
  4 meq/L for every 10 mmHg decrease in pCO_2

Case 1: Compensation in Acute Respiratory Acidosis:

A twenty-four-year-old male comes to the emergency room with a history of asthma and an acute asthma attack for the last four hours. He has not responded to albuterol that he uses inhaled. You obtain an ABG on this patient, which shows a pH of 7.29, a pCO_2 of 75, and a pO_2 of 65. The patient's electrolyte shows a sodium of 139, a potassium of 3.6, a chloride of 100, and a bicarbonate of 28. You realize that the patient has a respiratory acidosis. But, is this respiratory acidosis adequately compensated?

You realize that it is a respiratory acidosis since the pH is less than 7.40 and the pCO_2 is elevated, which would explain this alteration of the pH. The next step is to determine if this is acute or chronic. From the case you have been told that the patient has been having shortness of breath for only a few hours, so this is an acute respiratory acidosis which has not had enough time to compensate completely. So the formula for the expected bicarbonate on this patient is as follows:

HCO_3 increases by 1 meq/L for every 10 increase in the pCO_2. Since the pCO_2 in this patient has increased by 35 mmHg, the HCO_3 should increase by 3.5 or go from 24 to 27.5. The actual bicarbonate in this patient is 28, which is pretty close to 27.5. This indicates that the compensatory mechanism is adequate for this patient and there is no other metabolic disorder added on to this patient.

Now let's look at this case:

A twenty-four-year-old male comes to the emergency room with a history of asthma and an acute asthma attack for the last four hours. He has not responded to albuterol that he uses inhaled. You obtain an ABG on this patient, which shows a pH of 7.29, a pCO$_2$ of 75, and a pO$_2$ of 65. The patient's electrolyte shows a sodium of 139, a potassium of 3.6, a chloride of 100, and a bicarbonate of 34. You realize that the patient has a respiratory acidosis. But, is this respiratory acidosis adequately compensated?

In this case the expected bicarbonate would have been 27.5, but it is 34. This indicates that added to the respiratory acidosis, this patient also has a metabolic alkalosis.

Now, if you look at this case below:

A twenty-four-year-old male comes to the emergency room with a history of asthma and an acute asthma attack for the last four hours. He has not responded to albuterol that he uses inhaled. You obtain an ABG on this patient, which shows a pH of 7.29, a pCO$_2$ of 75, and a pO$_2$ of 65. The patient's electrolyte shows a sodium of 139, a potassium of 3.6, a chloride of 100, and a bicarbonate of 24. You realize that the patient has a respiratory acidosis. But, is this respiratory acidosis adequately compensated?

You expect to have a bicarbonate of 27.5 on this patient and the bicarbonate is actually 24 which is close to normal. It would look like there has been no compensatory mechanism at all, which is not the case. Actually, the problem is that the bicarbonate is less than expected, indicating that this patient has a respiratory acidosis with a metabolic acidosis added on to the problem. You can then see how to further assess metabolic acidosis by calculating the anion gap on the patient. In this case, the anion gap is 15, indicating that this is likely a high anion gap metabolic acidosis.

Case 2: Compensation in Chronic Respiratory Acidosis:

A sixty-four-year-old man comes to your office for a routine visit. The patient has a history of COPD for which he is on ipratropium bromide, albuterol, and theophilline. You obtain arterial blood gases which were reported to you as pH of 7.35, pCO$_2$ of 60, and pO$_2$ of 62 mmHg. You also obtain electrolytes which are as follows: sodium of 140, potassium of 3.8, a chloride of 100, and bicarbonate of 31. What is the acid-base disorder in this patient?

You realize that it is a respiratory acidosis since the pH is less than 7.40. You also notice that the pCO_2 is greater than 40 making this a primary respiratory acidosis. Now you have to make sure that it is adequately compensated. The compensatory formula is HCO_3 increases by 3.5 meq/L for every 10 mmHg increase in pCO_2. Since the pCO_2 has increased from 40 to 60 by a total of 20, 20 divided by 10 is 2. Multiply that by 3.5 and you get a total of 7. So, the bicarbonate should increase to 24 plus 7, which is 31. In this case, the bicarbonate has only increased to 31 which is adequately compensated. Therefore, we conclude that this is just a compensated respiratory acidosis.

Now if this case was as follows:

A sixty-four-year-old man comes to your office for a routine visit. The patient has a history of COPD for which he is on ipratropim bromide, albuterol, and theophilline. You obtain arterial blood gases which were reported to you as pH of 7.35, pCO$_2$ of 60, and pO2 of 62 mmHg. You also obtain electrolytes which are as follows: sodium of 140, potassium of 3.8, a chloride of 100, and bicarbonate of 36. What is the acid-base disorder in this patient?

In this case the primary disorder again is a respiratory acidosis, but the compensatory mechanism is supposed to be 31 as in the previous case. In this case, the bicarbonate is equal to 36 greater than what is expected, which must be due to an added metabolic alkalosis. So, the acid-base disturbance in this case is a respiratory acidosis with a metabolic alkalosis.

Now if this case was as follows:

A sixty-four-year-old man comes to your office for a routine visit. The patient has a history of COPD for which he is on ipratropim bromide, albuterol, and theophilline. You obtain arterial blood gases which are reported to you as pH of 7.35, pCO$_2$ of 60, and pO$_2$ of 62 mmHg. You also obtain electrolytes which are as follows: sodium of 140, potassium of 3.8, a chloride of 100, and bicarbonate of 26. What is the acid-base disorder in this patient?

In this case the primary disorder is again a respiratory acidosis. The compensatory mechanism is supposed to be 31 as previously stated, but in this case the bicarbonate is only 26 which indicates that the patient has a metabolic acidosis along with the respiratory acidosis. By calculating the anion gap in this patient, you can also see that it is 14, which would indicate that the patient has a high anion gap metabolic acidosis on top of his respiratory acidosis.

Case 3: Compensation in Acute Respiratory Alkalosis:

A twenty-five-year-old man presents to the emergency room with right flank pain for the last one hour. Urinalysis demonstrates that the patient has blood in the urine and you suspect that the patient is having a kidney stone, which is confirmed with an ultrasound of the kidneys. You obtain arterial blood gases since the patient's respiratory rate is elevated. It shows pH of 7.50, a pCO$_2$ of 25, and a pO$_2$ of 100, and his electrolytes show a sodium of 138, a potassium of 4.5, a chloride of 110, and a bicarbonate of 21. What is acid-base disorder in this patient?

You realize that it is a respiratory alkalosis since the pH is greater than 7.40 and the pCO$_2$ is low, which indicates that the primary disorder is an alkalosis. Since a low pCO$_2$ would cause an alkalosis, the primary disorder is a respiratory alkalosis. Now, the next step in a primary respiratory disturbance should be to establish if it is acute or chronic. In this case, the patient has had flank pain for one hour, so his problem is acute (only hours). The compensatory formula for acute respiratory alkalosis is that the HCO$_3$ reduces by 2 for every 10 decrease in the pCO$_2$. This patient's pCO$_2$ has gone down from 40 to 25 or 15 which would make the HCO$_3$ reduce by 1.5*2 or 3 or to 21. Because HCO$_3$ is also 21, it indicates that this patient has a compensated acute respiratory alkalosis.

Now if this was the case:

A twenty-five-year-old man presents to the emergency room with right flank pain for the last one hour. Urinalysis demonstrates that the patient has blood in the urine and you suspect that the patient is having a kidney stone, which is confirmed with an ultrasound of the kidneys. You obtain arterial blood gases since the patient's respiratory rate is elevated. It shows pH of 7.50, a pCO$_2$ of 25, and a pO$_2$ of 100, and his electrolytes show a sodium of 138, a potassium of 4.5, a chloride of 110, and a bicarbonate of 26. What is the acid-base disorder in this patient?

In this case we see that there is a primary respiratory alkalosis which should have a HCO$_3$ of about 21. But, when we see the bicarbonate of 26, we can see it is much higher than 21 indicating that there is also a metabolic alkalosis accompanying the primary respiratory alkalosis.

If that case was presented in this manner:

A twenty-five-year-old man presents to the emergency room with right flank pain for the last one hour. A urinalysis demonstrates that the patient has blood in the urine and you suspect that the patient is having a kidney stone, which is confirmed with an ultrasound of the kidneys. You obtain arterial blood gases since the patient's respiratory rate is elevated. It shows pH of 7.50, a pCO₂ of 25, and a pO₂ of 100, and his electrolytes show a sodium of 138, a potassium of 4.5, a chloride of 110, and a bicarbonate of 17. What is acid-base disorder in this patient?

As you can see in this primary respiratory alkalosis, the compensatory mechanism should be with a bicarbonate of 21. This bicarbonate is lower indicating that besides the respiratory alkalosis this patient also has a metabolic acidosis. We calculate the anion gap on this patient, which is 11 which is within normal. So, this is a non-anion gap metabolic acidosis.

Case 4: Compensation in Chronic Respiratory Alkalosis:

A thirty-two-year-old female on her eight month of pregnancy presents with difficulty in breathing for the last week. You obtain arterial blood gases which show a pH of 7.52, the pCO₂ as 23, and a pO₂ of 99. The patient's electrolyte shows a sodium of 148, a potassium of 3.4, a chloride of 95, and a bicarbonate of 17. What is the acid-base disorder in this patient?

You realize that it is a respiratory alkalosis since the pH is greater than 7.40, which can be explained by a low pCO₂ of 23. Next, we have to determine if this respiratory disorder is acute or chronic and we can see that it is chronic. For the last week, this woman has been having difficulty in breathing which is likely due to her pregnancy. So, the compensatory mechanism for the bicarbonate should be as follows: there should be a decrease of 4 meq/L for every decrease of 10 in the pCO₂. In this case the pCO₂ has decreased from 40 to 23 or by a total of 17. Therefore, we have to do the following math: 1.7 * 4 which is equal to 6.8. Therefore, the bicarbonate should decrease by 6.8, or 24 minus 6.8 equals 17.2 which is very close to 17. Therefore, this is a completely compensated chronic respiratory alkalosis.

Now, if this case was presented in the following manner:

A thirty-two-year-old female on her eight month of pregnancy presents with difficulty in breathing for the last one week. You obtain arterial blood gases which show a pH of 7.52, the pCO$_2$ as 2.3 and a pO$_2$ of 99. The patient's electrolyte shows a sodium of 148, a potassium of 3.4, a chloride of 95, and a bicarbonate of 22. What is acid-base disorder in this patient?

Again, this is a chronic respiratory alkalosis and the compensatory mechanism by the bicarbonate should have been 17. But, the bicarbonate in this case is higher, indicating that there is a metabolic alkalosis added on to the primary respiratory alkalosis.

Now, if this case was presented in the following manner:

A thirty-two-year-old female on her eight month of pregnancy presents with difficulty in breathing for the last one week. You obtain arterial blood gases which show a pH of 7.52, the pCO$_2$ as 23, and a pO$_2$ of 99. The patient's electrolyte shows a sodium of 139, a potassium of 3.4, a chloride of 106, and a bicarbonate of 14. What is the acid-base disorder in this patient?

Again, this is a chronic respiratory alkalosis and the compensatory mechanism should have brought the bicarbonate to 17. But, the bicarbonate is 14, which indicates that there must be a metabolic acidosis added on to this respiratory alkalosis. Then, if we calculate the anion gap we notice that it is 19 indicating that this is a high anion gap metabolic acidosis.

Triple Acid-Base Disturbances

So far you learned to identify the primary acid-base disturbance by looking at the pH, as well as, the pCO_2 and the bicarbonate, and also to identify double acid-base disturbances. Now it is time to be able to identify triple acid-base disturbance. This is a little more tricky and not as easy. You can have a metabolic acidosis with a metabolic alkalosis and a respiratory disorder, either an acidosis or alkalosis. Fortunately, since we cannot breathe fast and slow at the same time we never have a combined respiratory acidosis and alkalosis at the same time EVER. Or you can have a high anion gap metabolic acidosis with a normal anion gap metabolic acidosis and a respiratory disorder as well. In order to identify a triple acid-base disturbance it is always important to calculate the anion gap even when a patient has a primary metabolic alkalosis.

The third step in analyzing an acid-base disturbance is to calculate the anion gap. Normally, in a metabolic alkalosis the anion gap is slightly elevated, but never more than five above what it normally would be. That is, the anion gap in a metabolic alkalosis is usually never more than 17. If you calculate the anion gap in a patient with a metabolic alkalosis and it is greater than 17, then there is also a hidden high anion gap metabolic acidosis in the patient.

In summary, the steps in analyzing an acid-base disturbance are:

a) Determine the primary acid-base disturbance: This is done by looking at the pH. If the pH is less than 7.40 it is an acidosis and if it is greater than 7.40 it is an alkalosis. Then, by looking at the pCO_2 and the bicarbonate you can determine if the pCO_2 or the bicarbonate can explain the acid-base disturbance, to see if it is a primary respiratory or metabolic disorder.

b) b) Determine the compensatory mechanism: This is done simply by looking at the last two digits of the pH and the pCO_2 in primary metabolic disorders, but it is slightly more complex in respiratory disorders where the compensatory mechanism by the bicarbonate depends on whether the disorder is acute or chronic.

c) c) Always calculate the anion gap: The anion gap is calculated by the following formula.

$$Anion\ Gap = Na - (HCO_3 + Cl)$$

Normally, the anion gap is around 8 to 12. If the anion gap is greater than 12 in a metabolic acidosis it is considered a high anion gap metabolic acidosis. In a metabolic alkalosis the anion gap can be as high as 17, but if it is higher than 17 it indicates that there is a hidden high anion gap metabolic acidosis as well in the patient. So, it becomes very important to identify a triple disturbance to evaluate the anion gap even in a metabolic alkalosis, which will be when you realize that there is both a metabolic alkalosis with a metabolic acidosis and by evaluating the compensatory mechanism you can determine if there is also a respiratory disorder.

I will give a few examples of triple acid-base disturbances.

Case 1: Triple Acid-Base Disorder:

A patient admitted to the intensive care unit of a hospital has the following arterial blood gases which show a pH of 7.59, a pCO_2 of 50, and a pO_2 of 75. The patient's electrolyte shows a sodium of 145, a potassium of 2.8, a chloride of 91, and a bicarbonate of 28. What is the acid-base disorder in this patient?

As you can see in this case, the patient has a primary alkalosis, and because the pCO_2 is greater than 40 it is likely to be a metabolic alkalosis. When we look at the bicarbonate it confirms our suspicion since it is 28. If this would have been a respiratory alkalosis the pCO_2 would have been less than 40, and that is not the case in this patient. Then, when we see if this is a compensated metabolic alkalosis we would have expected the pCO_2 to be equal to (28-24) * 0.7 + 40 (measured bicarbonate minus normal bicarbonate time 0.7 plus 40 equals the expected compensatory mechanism in a metabolic alkalosis) this should be around 42.8 and it is actually much higher, indicating that this patient has an added respiratory acidosis as well. Now, when we calculate the anion gap on this patient we see that it is (145-(91+28))= 26, which is much higher than 17, indicating that there is also a hidden high anion gap metabolic acidosis. Therefore, this patient has a triple acid-base disturbance, which is a metabolic alkalosis (the primary disturbance), a respiratory acidosis, and a high anion gap metabolic acidosis.

Case 2: Triple Acid-Base Disorder:

A patient admitted to the intensive care unit of a hospital has the following arterial blood gases which show a pH of 7.21, a pCO$_2$ of 35, and a pO$_2$ of 85. The patient's electrolytes show a sodium of 140, a potassium of 3.7, a chloride of 110, and a bicarbonate of 15. What is the acid-base disorder in this patient?

As you can see, in this case the patient has a primary high anion gap metabolic acidosis. The compensatory mechanism of a metabolic alkalosis is obtained by multiplying 0.5 by the measured bicarbonate and adding eight plus or minus 2, this would be equal to (15*1.5) + 8 plus or minus 2, which is between 28.5 and 32.5 and in this patient the pCO2 is much higher than that indicating that the patient also has an added respiratory acidosis with this disorder, notice than in triple acid base disturbances you cannot utilize the previous trick where the you can compare the last 2 digits of the pH with the pCO2 since this could be misleading due to the third acid base disturbance which will be contributing to the pH as well. Now, this patient can also have an added normal anion gap metabolic acidosis or a metabolic alkalosis. This can be figured out by calculating the delta-delta ratio which is basically done with the following formula:

(Change in anion gap from normal) divided by (Change in bicarbonate from normal)

In this case the anion gap is 15 and normally should be 12; so, the change in the anion gap is 3. The change in the bicarbonate is calculated by subtracting from the normal bicarbonate of 24, the measured bicarbonate which is 15, which gives you 9. So if you divide 3/9 it gives you a number less than one or 0.333. Normally, this delta-delta ratio is 1 or 2. If the delta-delta ratio is less than one like in this patient, it is because the patient also has a normal anion gap metabolic acidosis hidden as well as the high anion gap metabolic acidosis, plus the respiratory acidosis. Therefore, this patient has a triple acidosis.

By knowing this simple acid-base approach you can calculate almost any acid-base disturbance that a patient might present with, including the most complicated of cases and triple acid-base disturbances without much difficulty.

Metabolic Acidosis Assessment

Let's evaluate metabolic acidosis at this time. As previously discussed metabolic acidosis is suspected when the pH is low or less than 7.40, and the patient's bicarbonate is also low at usually less than 20. If the bicarbonate is less than 15, the patient definitely has a metabolic acidosis. Also, the respiratory system compensates for a metabolic acidosis with a respiratory alkalosis by breathing fast and usually bringing the pCO_2 down. The next step in the assessment of a metabolic acidosis is to calculate the anion gap with the following formula:

$$Anion\ Gap = Na - (HCO_3 + Cl)$$

Normally, the anion gap is less than 12, usually between 8 and 12. If the anion gap is greater than 12 then it is considered a high anion gap metabolic acidosis. If the anion gap is normal it is usually a normal anion gap metabolic acidosis, where most of the renal tubular acidosis falls, diarrhea and other diagnosis as well.

Assessment of High Anion Gap Metabolic Acidosis

The next step in identifying a high anion gap metabolic acidosis is to calculate the plasma osmolal gap in the following manner. Usually, the osmolal gap is calculated by obtaining the serum osmolality from the laboratory and by calculating the serum osmolality with the following formula:

$$Posm = 2[Na+] + 10$$

If the glucose and the BUN are within normal limits. Otherwise, you would have to use the following formula:

$$Posm = 2[Na+] + Glucose/18 + BUN/2.8$$

You can calculate the plasma osmolal gap by figuring out the change in the reported plasma osmolality by the laboratory and the calculated plasma osmolality that you calculated. If this is greater than 15 it is considered a high plasma osmolal gap. Two conditions you should suspect if the plasma osmolal gap is high, it is going to be intoxication with 1) methanol and 2) ethylene glycol.

Patients who have a metabolic acidosis due to methanol will have a high anion gap metabolic acidosis. With a high osmolal gap, these patients also complain of visual disturbance.

Patients with ethylene glycol ingestion will also have a high anion gap metabolic acidosis with an elevated osmolal gap. If you check the urine sediment, you will see abundant oxalate crystals in the urine, which will give you a clue to this diagnosis.

Other causes of metabolic acidosis which are high anion gaps can be suspected from the clinical presentation in the patient such as diabetes, where the patient's blood glucose will be elevated and he or she will also have positive serum ketones. The patient will have positive serum ketones in alcoholic ketoacidosis which also causes a high anion gap metabolic acidosis In uremia, the patient's BUN and creatinine would be elevated, and in hypoxemia the patient will have lactic acid positive, or there would be a medication which can cause lactic acidosis such as metformin.

Assessment of Normal Anion Gap Metabolic Acidosis

Case 1: Normal anion gap metabolic acidosis:

> This is a forty-five-year-old male with type 1 diabetes, who is on a transplant list at a local hospital due to end stage renal disease. His glycosilated hemoglobin is 5.7, his BUN is 5.5 and creatinine is 4.11. The patient's sodium is 138, potassium 5.3, he is currently on kayaxalate, his chloride is 109, and a bicarbonate is 18. What is the most likely diagnosis in this patient?
>
>   a) Uremic acidosis
>   b) Diabetes ketoacidosis
>   c) Renal tubular acidosis type 1
>   d) Renal tubular acidosis type 2
>   e) Renal tubular acidosis type 4

When doing the assessment of a normal anion gap metabolic acidosis you should be able to think of the normal anion gap metabolic acidosis and consider the possible diagnosis and how to determine the most likely diagnosis. First thing you should think of is the renal tubular acidosis or the possibility that it is due to gastrointestinal fluid loss. The patient in our example does not have any clinical evidence of gastrointestinal fluid loss such as diarrhea and that is not an option given. We know that options a and b both cause a high anion gap metabolic acidosis and this is definitely a normal anion gap metabolic acidosis. That leaves us with the option of renal tubular acidosis. As we can see, the most common are type 1, type 2, and type 4. In this case the most likely diagnosis is type 4 renal tubular acidosis because this patient's potassium is in the upper limit of normal, even after taking kayaxalate which is used to lower potassium in patients with hyperkalemia. Type 1 and type 2 renal tubular acidosis are not associated with hyperkalemia, but mostly with hypokalemia. The next step in assessing a patient with a normal anion gap metabolic acidosis is to calculate the urine anion gap, which is done with the following formula:

$$\text{Urine AG} = \text{Urine } (Na + K - Cl).$$

A negative urine anion gap indicates excess in ammonium excretion which is not measured and indicates that the kidneys are excreting acid appropriately; therefore, the problem would not be related to the kidneys. If the urine anion gap is negative the cause of the normal anion gap metabolic acidosis is due to a gastrointestinal cause, whereas a positive urinary anion gap will indicate a renal tubular acidosis.

The next thing that you should do is to obtain a urinary pH, the urinary pH greater than 5.4 or 5.5 is seen in patients with type 1 renal tubular acidosis or distal renal tubular acidosis, where the defect is due to lack of acidification of the urine in the distal tubules. If the urinary pH is less than 5.4 it could be due to gastrointestinal fluid loss or due to a type 2 or type 4 renal tubular acidosis.

In the next step, if you still suspect a renal tubular acidosis, look at the serum potassium. If it is elevated, it would be a type 4 renal tubular acidosis. It would be a type 2 renal tubular acidosis, if the urinary pH is less than 5.4 and the potassium is normal or low.

Metabolic Alkalosis Assessment

This is the simplest of the acid-base disturbances. You recognize a metabolic alkalosis on a patient with 1) a high pH or a pH greater than 7.40 and a high bicarbonate; 2) a respiratory disturbance and the compensatory mechanism showing that the bicarbonate is higher than expected; or 3) a high anion gap metabolic acidosis with a delta-delta ratio greater than 2. The most common causes of metabolic alkalosis is either due to vomiting or nasogastric suction or an equivalent condition to this or due to diuretic use or excess renal excretion of hydrogen ions. These two conditions correspond to about 90% of the cases of patients with metabolic alkalosis. Excess mineralocorticoid also causes metabolic alkalosis in patients with conditions where there is excess aldosterone. In such cases usually, the patient will also have hypokalemia and increase in blood pressure with increase in fluid retention. The blood pressure is usually decreased in patients who are vomiting or who are taking diuretic due to the volume contraction that occurs in them.

The first step in the management of a metabolic alkalosis is to obtain a urine electrolyte to determine the value of the urine chloride. If it is greater than 20, it is likely to be due to diuretic use, as well as, excessive mineralocorticoid effects. If it is low it is most likely a chloride responsive metabolic alkalosis and usually, it is treated with normal saline replacement. It is important to check the blood pressure of the patient with a metabolic alkalosis since a high blood pressure indicates increase in aldosterone such as in patient with primary hyperaldosteronism, which usually also has an elevated chloride in the urine. But in patients using diuretic, even though the chloride in the urine is high their blood pressure tends to be normal or even low.

Patients with a chloride responsive metabolic alkalosis are generally treated with normal saline. If the patient is fluid overloaded, normal saline will be detrimental to the patient, and in those cases patients should be treated with acetazolamide, which will help excrete excess bicarbonate in the urine improving the metabolic alkalosis.

Respiratory Acidosis Assessment

Respiratory acidosis is usually caused by hypoventilation due to neurological conditions, drugs or any condition which decreases the stimulation for breathing. Usually patients with this type of respiratory acidosis have a slow respiratory rate. Sometimes the cause of the respiratory acidosis is lack of gas interchange in the lungs and in these cases this type of respiratory acidosis is accompanied by a fast respiratory rate and it is clear that these patients are in impending respiratory failure. Breathing fast will always cause a decrease in the pCO_2, which will cause a respiratory alkalosis. Therefore, if anyone is breathing fast and has an elevated pCO_2 or a respiratory acidosis it is due to lack of exchange of gases in the lungs and indicates possible impending respiratory failure. The airways of these patients must be protected promptly by either noninvasive positive pressure ventilation or intubation and mechanical ventilation.

Respiratory Alkalosis Assessment

Respiratory alkalosis is usually caused by conditions that increase the rate at which a patient breathes. Conditions which can acutely cause a patient to breath fast are anxiety and acute onset of pain. More chronic conditions are seen, for example, in the late stages of pregnancy due to abdominal distention, and for that matter any condition which causes abdominal distention will lead the patient to breath faster than normal, and this could lead to respiratory alkalosis.

As with respiratory acidosis, the compensatory mechanism takes place by the bicarbonate system which will take some time, usually days, for this to occur fully. The kidneys will try to eliminate more bicarbonate in the proximal tubules to compensate for the alkalosis, decreasing the serum bicarbonate somewhat as we saw previously.

Practice Questions

1. A twenty-three-year-old college student returned from an archeological trip in Guatemala a day ago. The patient has generally been in good health. He spent one week with a family in Guatemala in a remote area. The patient started to develop some loose bowel movements the day before returning to the United States. Currently, the patient is experiencing voluminous watery diarrhea with no fever, blood in the stools or abnormal skin lesions. He states he has been consuming large amounts of Gatorade to keep himself hydrated, but since he has been feeling somewhat weak, he was concerned and decided to come for further evaluation. On physical examination, the patient's blood pressure is 110/70 and is not orthostatic, his respiratory rate is 19, he does not have a fever and his heart rate is 92 beats per minute. Laboratory reports obtained on the patient show that his sodium is 132 meq/L, potassium 3.5 meq/L, chloride 105 meq/L, and bicarbonate 18 meq/L. Urine obtained from the patient shows a pH of 5.9, urine electrolytes are reported as a urine sodium of 30, a urine potassium of 32, and a urine chloride of 55. Arterial blood gases while breathing room air are as follows: pH of 7.35, pCO_2 of 34 mmHg, and pO_2 of 94 mmHg. Which of the following is the most likely cause of this patient's acid-base disturbance?

   a. Travelers' diarrhea
   b. Lactic acidosis
   c. Type 4 renal tubular acidosis
   d. Type 1 renal tubular acidosis
   e. E. coli O157:H7

2. A fifty-two-year-old woman with type 2 diabetes presents to the hospital complaining of diarrhea, abdominal discomfort for the last few days. The patient is currently taking metformin to control her diabetes, and uses enalapril to control hypertension. The patient is complaining of difficulty breathing. On physical examination, her temperature is 37.2 °C, her pulse is 98, respiration rate 21, and her blood pressure is 105/65. Laboratory reports obtained on the patient show that her sodium is 139 meq/L, potassium 3.2 meq/L, chloride 115 meq/L, and bicarbonate 16 meq/L. Urine obtained from the patient shows a pH of 5.3, urine electrolytes are reported as urine sodium of 25, urine potassium of 22, and a urine chloride of 75. Arterial blood gases while breathing room air are as follows: pH of 7.32, pCO_2 of 33 mmHg, and pO_2 of 90 mmHg. Which of the following is the most likely cause of this patient's acid-base disturbance?

    a. Diabetes ketoacidosis
    b. Lactic acidosis
    c. Type 2 renal tubular acidosis
    d. Type 1 renal tubular acidosis
    e. Diarrhea

3. A sixty-five-year-old man with type 2 diabetes being treated with glyburide and metformin presents to the emergency room due to weakness and slight disorientation. His son tells you that his father likes to drink alcohol once in a while and that he does not follow the diet for his diabetes as his doctor had told him. He also mentions that his father had been complaining of diarrhea for the last few weeks. On physical examination, the patient's temperature is 37.8 °C, his pulse 88, respiration rate 17, and his blood pressure 135/95. Laboratory reports obtained on the patient show that his sodium is 135 meq/L, potassium 5.9 meq/L, chloride 107 meq/L, and bicarbonate 17 meq/L. Urine obtained from the patient shows a pH of 5.4, urine electrolytes are reported as urine sodium of 28, urine potassium of 32, and a urine chloride of 55. Arterial blood gases while breathing room air are as follows: pH of 7.33, pCO_2 of 34 mmHg, and pO_2 of 87 mmHg. Which of the following is the most likely cause of this patient's acid-base disturbance?

    a. Diarrhea
    b. Alcoholic ketoacidosis

   c. Type 2 renal tubular acidosis
   d. Type 1 renal tubular acidosis
   e. Type 4 renal tubular acidosis

4. A fifty-four-year-old woman admitted to the intensive care unit on a mechanical ventilator was brought to the hospital by paramedics. The patient was found unconscious on the street and passersby called 911. On physical examination, her temperature was 36.5 °C, her pulse was 55, respiration rate was 10, and her blood pressure was 125/75. Laboratory reports obtained on the patient showed that her sodium was 142 meq/L, her potassium was 2.8 meq/L, her chloride was 88 meq/L, and her bicarbonate was 12 meq/L. Arterial blood gases while breathing room air are as follows: pH of 7.26, pCO_2 of 35 mmHg, and pO_2 of 99 mmHg on 80% FiO_2. Which of the following best describes this patient's acid-base disorder?

   a. Mixed high anion gap metabolic acidosis/respiratory alkalosis
   b. Mixed normal anion gap metabolic acidosis/metabolic alkalosis/respiratory alkalosis
   c. Mixed high anion gap metabolic acidosis/metabolic alkalosis/respiratory alkalosis
   d. Mixed high anion gap metabolic acidosis/metabolic alkalosis/respiratory acidosis
   e. Mixed high anion gap metabolic acidosis/respiratory acidosis

5. A thirty-four-year-old patient with AIDS who has not been compliant with his antiretroviral regimen is admitted to a hospital with cryptococcal meningitis, pneumocistis jiroveci pneumonia, as well as HIV wasting syndrome. He is put on a mechanical ventilator and on total parenteral nutrition. On physical examination, his temperature is 38.8 C, his pulse is 112, respiration rate is 12, and his blood pressure is 105/60. Laboratory reports obtained on the patient show that his sodium is 137 meq/L, potassium 4.3 meq/L, chloride 110 meq/L, and bicarbonate 10 meq/L. Urine obtained from the patient shows a pH of 5.8, urine electrolytes are reported as urine sodium of 17, urine potassium of 22, and a urine chloride of 25. Arterial blood gases while breathing room air are as follows: pH of 7.23, pCO_2 of 24 mmHg, and pO_2 of 82 mmHg. Which of the following is the most likely cause of this patient's acid-base disturbance?

a. Mixed high anion gap metabolic acidosis/respiratory alkalosis
b. Mixed high anion gap metabolic acidosis/RTA type 1
c. Mixed high anion gap metabolic acidosis/normal anion gap metabolic acidosis/respiratory alkalosis
d. Mixed high anion gap metabolic acidosis/RTA type 2
e. Mixed high anion gap metabolic acidosis/normal anion gap metabolic acidosis due to a non-renal cause of acidosis

6. A sixty-two-year-old female with COPD presents to you for a routine follow-up visit. She has a fifty-pack-year history of smoking. You obtain the following routine blood work including electrolytes reported as follows: sodium of 142, potassium of 3.4, chloride of 105, and bicarbonate of 28. Arterial blood gases on room air are reported as follows: pH of 7.34, pCO_2 of 65, and pO_2 of 55. Which of the following is the acid-base disturbance present in this patient?

a. Respiratory acidosis and a metabolic alkalosis
b. Respiratory acidosis with a respiratory alkalosis
c. Respiratory acidosis fully compensated
d. Metabolic acidosis with a respiratory alkalosis
e. Respiratory acidosis and a metabolic acidosis

7. A thirty-two-year-old male is seen in the emergency room of a hospital due to left flank pain which started about an hour ago. The patient has had a history of nephrolithiasis in the past and informs you that the pain he has resembles the pain from his prior kidney stone which he passed about one year ago. A urine dipstick shows 3+ blood. You obtain the following blood work: a sodium of 140, a potassium of 3.6, a chloride of 110, and a bicarbonate of 19. Urine pH is 5.7, urine sodium is 20, urine potassium is 25, and the urine chloride is 30. Arterial blood gases obtained show a pH of 7.49, a pCO_2 of 30, and a pO_2 of 99. What is the most likely acid-base disturbance?

a. Respiratory alkalosis and a metabolic alkalosis
b. Respiratory alkalosis with a RTA type 1
c. Respiratory alkalosis with a RTA type 2
d. Metabolic alkalosis with a RTA type 4
e. Metabolic alkalosis with respiratory alkalosis

8. A twenty-three-year-old female with history of bulimia presents to a hospital with altered mental status. The patient looks chronically ill. You obtain the following laboratory reports on this patient: a sodium of 140, a potassium of 3.6, a chloride of 92, and a bicarbonate of 18. Arterial blood gases obtained on the patient showed a pH of 7.34, a pCO_2 of 35, and a pO_2 of 87. Which of the following best explains this patient's acid-base disturbance?

    a. Respiratory acidosis
    b. High anion gap metabolic acidosis
    c. Normal anion gap metabolic acidosis
    d. Mixed high anion gap metabolic acidosis with a metabolic alkalosis
    e. Mixed normal anion gap metabolic acidosis with a respiratory acidosis

9. A thirty-four-year-old man with type 1 diabetes presents to your office for a routine evaluation complaining of excessive fatigue. You obtain the following blood work: BUN of 42, creatinine of 3.8, sodium of 138, potassium of 6.1, chloride of 110, and bicarbonate of 17. What is the most likely diagnosis?

    a. Respiratory alkalosis
    b. Diabetes ketoacidosis
    c. Uremia
    d. Type 1 RTA
    e. Type 4 RTA

Answers to Practice Questions

1. The correct answer for this question is type 1 renal tubular acidosis. This patient has a pH which is less than 7.40 indicating that the primary disorder must be an acidosis. The pCO_2, which is less than 40 and the bicarbonate, which is low indicate that the primary disturbance is a metabolic acidosis. When we calculate the anion gap in this patient we notice that it is 9 which is a normal anion gap metabolic acidosis. The next step in evaluating this acid-base disturbance is to determine if the kidneys are appropriately handling the acidosis by excreting appropriate amount of ammonia. This can be done by calculating urine anion gap (USodium + UPotassium - UChloride), which gives a positive value indicating that the kidney is not excreting appropriate amount of ammonia, which would be the case in patients with a positive urine anion gap. This tells us that even though this patient has diarrhea it is unlikely for this to be the cause of his acidosis, because in that case his urine anion gap would have been a negative number since ammonia is not measured in the urine. It must be a renal tubular acidosis causing his problem. The next step in evaluating renal tubular acidosis is to look at the urinary pH. If it is greater than 5.4 or 5.5 it is usually a type 1 or distal renal tubular acidosis, since this is the type of renal tubular acidosis which does not allow the kidneys to acidify the urine appropriately to a level of 5.5 or less. Therefore, the correct answer is type 1 renal tubular acidosis.

    If the urine anion gap was negative then travelers' diarrhea would have been the correct choice. Type 4 renal tubular acidosis usually has a normal urinary pH and the serum potassium in these patients would have been high, and not low or normal like this patient has. Lactic acidosis causes a high anion gap metabolic acidosis and this patient's metabolic acidosis has a normal anion gap. Finally, E. coli O157:H7 causes an inflammatory type of diarrhea with usually blood in the stools. This is evidence of a microangiopathic hemolytic anemia, where there would have been mention of anemia with schistocytes in the peripheral blood smear, elevated LDH, petechia, and perhaps low platelet which this patient did not appear to have, or the information provided did not reveal any evidence of an inflammatory diarrhea indicating the possibility of E. coli O157:H7.

2. The correct answer for this question is diarrhea. The first thing you do is determine the primary disturbance in this patient. This is done by looking at the pH which is less than 7.40 indicating an acidosis. The pCO_2 which is less than 40 will not explain the acidosis, and looking at the bicarbonate you notice that it is low indicating that the primary disturbance is a metabolic acidosis. Then you calculate the anion gap and it is 8 indicating a normal anion gap metabolic acidosis. Now, the next challenge here is to determine if this is due to a gastrointestinal fluid loss or due to a renal cause such as a renal tubular acidosis. The urine anion gap, which in this patient is -28 indicates that the kidneys are eliminating appropriate amount of ammonia. Since the urine anion gap is negative, this cannot be due to a renal tubular acidosis. It must be due to an extrarenal cause of an acidosis like the diarrhea that the patient is having.

    Diabetes ketoacidosis is more common in type 1 diabetics. This patient has type 2 diabetes, and since diabetes ketoacidosis is a high anion gap metabolic acidosis this patient's anion gap is normal. Even though metformin can cause lactic acidosis, again lactic acidosis causes a high anion gap metabolic acidosis, which is not the case in this patient.

    The other two responses—type 1 or a type 2 renal tubular acidosis—cannot be correct in this patient because the urinary anion gap in those cases would have been a positive number and this urinary anion gap is -28, which is negative indicating that the kidneys have no problem handling this acidosis and is excreting increasing amount of ammonia which is not measured. Also, in type 1 renal tubular acidosis the urinary pH is usually above 5.4 and this urinary pH is less than 5.4 which makes a type 1 renal tubular acidosis even less likely.

3. The correct answer for this question would be type 4 renal tubular acidosis. The first thing is that the pH in the arterial blood gases of this patient is less than 7.40 indicating an acidosis, then you look at the pCO_2 and notice that it is low which cannot explain this acidosis. So the primary disorder is most likely a metabolic acidosis which when you see

the bicarbonate you notice that it is low or less than 24 indicating that this is likely a primary metabolic acidosis which is purely compensated specially when you look at the last two digits of the pH and the pCO_2 they are within one to two digits of each other. The next step is to determine if this acidosis is caused by a renal problem or by an extrarenal problem by calculating the urine anion gap, which if positive would indicate that the kidneys are not functioning appropriately indicating a renal tubular acidosis. In this case the urine anion gap is +5 which indicates that the kidneys are not excreting appropriate amount of ammonia. Therefore, this patient is most likely to have renal tubular acidosis. The next step is to check the urinary pH which in this patient is not greater than 5.4 indicating that this cannot be a type 1 or distal renal tubular acidosis where the urinary pH is usually greater than 5.4. Next, check the serum potassium because in a type 4 renal tubular acidosis the potassium is usually high like in this case. In a type 2 renal tubular acidosis the potassium is not elevated.

In this case alcoholic ketoacidosis would have caused a high anion gap metabolic acidosis, and the metabolic acidosis in this patient is a normal anion gap metabolic acidosis. Diarrhea is another option in this patient, but diarrhea which is the most common cause of a normal anion gap metabolic acidosis would not cause a problem in the excretion of ammonia by the kidneys. If this case would be due to diarrhea the patient's urine anion gap would have been a negative number indicating that the kidneys were excreting appropriate amount of ammonia which is not the case with this urine anion gap of +5.

4. The correct answer is a triple acid-base disturbance of high anion gap metabolic acidosis, a metabolic alkalosis, and a respiratory acidosis. The first step in analyzing this acid-base disturbance is to see what the primary disturbance is. In this case the pH being less than 7.40 is an acidosis and since the pCO_2 is less than 40 and the bicarbonate is low it is likely to be a metabolic acidosis. Then, you calculate the anion gap in this patient which is 42 indicating a high anion gap metabolic acidosis. Then we would like to see if this is appropriately compensated by the respiratory system. The compensatory mechanism should be a pCO_2 of around 26 give or take 2 and in this case the pCO_2 is actually 35 indicating that there must also be a respiratory acidosis because the pCO_2 is much higher than expected. Then, we can check the delta-delta ratio to see if there is another metabolic disturbance hidden in this patient. It can either be a non anion gap metabolic acidosis if the delta-delta

ratio is less than one or a metabolic alkalosis if the delta-delta ratio is more than 2. The calculation of the ratio is done by determining the change in the anion gap which in this case is (40 - 12 = 28) dividing that by the change in the bicarbonate which in this case is (24 - 12 = 12) and 28/12 is definitely greater than 2, indicating that there is a hidden metabolic alkalosis in this patient as well. If this ratio would have been less than one, then that would have indicated that the patient had a non anion gap metabolic acidosis added on to his high anion gap metabolic acidosis, but this is not the case here.

If you had not calculated the delta-delta ratio, you would have missed the metabolic alkalosis hidden in this patient, and probably your answer would have been the obvious high anion gap metabolic acidosis. When you checked the compensatory mechanism you would have also noted that there was a respiratory acidosis since the patient's pCO_2 is much higher than what is expected for this level of bicarbonate.

5. The correct answer for this case is a mixed high anion gap metabolic acidosis with a normal anion gap metabolic acidosis RTA type 1. First notice that this patient's pH is less than 7.40 indicating that the patient has an acidosis which can be explained by the low bicarbonate in this patient as a metabolic acidosis. Then, we calculate the anion gap and notice that it is high; therefore, it constitutes a high anion gap metabolic acidosis. Next we check if this metabolic acidosis is well compensated or not. We expect the last two digits of the pH to be equal or close to the pCO_2 and in this case it is true indicating that this metabolic acidosis is fully compensated by a respiratory alkalosis. Therefore, there is no added respiratory disorder. But, since this is a high anion gap metabolic acidosis you can calculate the delta-delta ratio which in this case would have been the change in the anion gap (17 - 12 = 5) divided by (24 - 10 = 14); therefore, 5 divided by 14 is much less than 1 which tells you there is also an added non anion gap metabolic acidosis, and by doing the urine anion gap you notice that the value is a positive value indicating that this patient likely has a renal tubular acidosis which is likely a type 1 renal tubular acidosis since the urine pH is greater than 5.4.

6. The correct answer here is a mixed respiratory acidosis with a non anion gap metabolic acidosis. As you can see from this case the primary disorder would be an acidosis since the pH in this patient is less than 7.40, and the pCO_2 is elevated which indicates that this is a primary respiratory acidosis. Then, by calculating the compensatory mechanism we realize

that this patient has a chronic condition, where the bicarbonate should increase by 4 mmHg for every 10 increase in the pCO_2. Since the pCO_2 has increased by 25 from 40, the bicarbonate should also increase by ($2.5 \times 4 = 10$) or to 34 but the bicarbonate on this patient is much lower than this indicating that this patient has a metabolic acidosis on top of respiratory acidosis. By checking the anion gap, we notice that it is 9 indicating that this is a non anion gap metabolic acidosis.

7. The best answer for this patient is a respiratory alkalosis with a RTA type 1. The pH which is greater than 7.40 indicates that the primary acid-base disturbance is an alkalosis, and the pCO_2 indicates that it is low indicating that it is likely a respiratory alkalosis. This is an acute respiratory alkalosis which started one hour ago. Therefore, for every 10 decrease in the pCO_2, the bicarbonate should also decrease by 1. Since the normal bicarbonate is usually 24 the bicarbonate should not be less than 23 and definitely not less than 20 like in this patient which indicates that this patient also has a metabolic acidosis on top of the respiratory alkalosis. Then calculating the anion gap we notice that the anion gap is within normal limits, therefore, the patient has a normal anion gap metabolic acidosis. By looking at the pH of the urine, it is greater than 5.4 which could imply a type 1 renal tubular acidosis. Then, calculating the urine anion gap, we notice that it has a positive value, which indicates a renal tubular acidosis likely a type 1 renal tubular acidosis with a respiratory alkalosis.

8. This patient's primary disturbance is a high anion gap metabolic acidosis and when we look at the pH it is 7.34 and the pCO_2 is 35. Therefore, it is a fully compensated metabolic acidosis since 34 and 35 are very close to each other. The next thing which should not be missed in a case like this is another metabolic acidosis or a hidden metabolic alkalosis. To do this we calculate the delta-delta ratio in this patient which is the difference in the anion gap. In this case the anion gap is 30 and normal anion gap is 12, so the difference is $30 - 12 = 18$, and divide that by the difference in the bicarbonate. In this case the bicarbonate is 18, so we get $24 - 18 = 6$. So $18/6 = 3$, which is greater than 2 indicates that there is also a hidden metabolic alkalosis which is likely secondary to the chronic vomiting in this patient.

9. This patient who is a type 1 diabetic is prone to diabetes ketoacidosis but when we calculate the anion gap in this patient it is only 11 which is a normal anion gap. Therefore, it is unlikely that the answer is diabetes ketoacidosis in this case. Now, if the patient has a respiratory alkalosis,

the patient would compensate with a metabolic acidosis, but there is mention of this patient breathing fast, which is an indication of a primary respiratory disturbance. Now, the laboratory reports show that the BUN and creatinine in this patient is elevated, so can this be Uremia causing this abnormality and acidosis. Well, uremia can cause an acidosis, but it is a high anion gap metabolic acidosis and this acidosis is a normal anion gap metabolic acidosis. So, we are left with either a type 1 or 4 RTA. Since the potassium is elevated, we know that the best diagnosis in this case would be type 4 RTA and not type 1 which does not present with an elevated potassium; in fact, usually the potassium is low.

Index

A

B

C

D